Simeon Snell

On the Prevention of Eye Accidents Occurring in Trades

An address delivered at the opening of the Section of Ophthalmology at

the annual meeting of the British Medical Association at Portsmouth,

August, 1899

Simeon Snell

On the Prevention of Eye Accidents Occurring in Trades
*An address delivered at the opening of the Section of Ophthalmology at the annual
meeting of the British Medical Association at Portsmouth, August, 1899*

ISBN/EAN: 9783337254629

Printed in Europe, USA, Canada, Australia, Japan

Cover: Foto ©berggeist007 / pixelio.de

More available books at **www.hansebooks.com**

On the Prevention of Eye Accidents Occurring in Trades

An Address Delivered at the Opening of the Section of Ophthalmology at the Annual Meeting of the British Medical Association at Portsmouth, August, 1899

By SIMEON SNELL, F.R.C.S.Edin.

Ophthalmic Surgeon to the Royal Infirmary, Sheffield ; and Professor of Ophthalmology, University College, Sheffield ; President of the Section.

London

JOHN BALE, SONS & DANIELSSON, Ltd.

OXFORD HOUSE

83-89, GREAT TITCHFIELD STREET, OXFORD STREET, W.

1899

ON THE PREVENTION OF EYE ACCIDENTS OCCURRING IN TRADES.

An Address delivered at the Opening of the Section of Ophthalmology, at the Annual Meeting of the British Medical Association at Portsmouth, August, 1899,

By SIMEON SNELL, F.R.C.S.Edin.

Ophthalmic Surgeon to the Royal Infirmary, Sheffield; and Professor of Ophthalmology, University College, Sheffield; President of the Section.

My first duty is the pleasing one of cordially acknowledging the honour conferred in selecting me to preside over the Section of Ophthalmology, and, whilst as a stranger, like many others amongst us, it ill becomes me to bid you welcome to this town, I can and do give a hearty welcome to all who may attend our meetings, and express the hope that they will freely take part in the discussions, and will also derive profit from our deliberations.

I desire to ask your kind attention for a short time to some remarks I propose to make on the very important subject of the causation and prevention of eye accidents in certain trades. It is not possible in the limited time at my disposal to consider at all completely such a very wide subject. Practically my whole professional life has been spent in a very populous district well known for its extensive iron and steel works, for its cutlery, and for a number of metal and other trades, and also as the centre of a very extensive coalfield. It is possible that eye accidents occur there with as great frequency as in any other district that could be named, and if the number of accidents is large the methods by which they are occasioned are also very varied. Experience gained in such a field gives one, perhaps, some title to speak of the causes of these accidents, and the means which should be adopted to obviate them.

No one can have seen much of ophthalmic

practice in such a large industrial centre without being painfully aware of the enormous destruction to sight occasioned annually by accident. It is difficult, of course, to obtain anything like accurate statistics. Magnus, in his tables, makes 8·5 per cent. of all cases of blindness as due to accident. Those blind in one eye only and the far larger number who have sustained permanent injury in varying degrees short of blindness are excluded in such a calculation, and, however true such a statement may be for the community generally, the number must be greatly exceeded in large and populous centres, especially those in which iron and steel are important industries.

STATISTICS.

I do not intend to burden my remarks with many statistics. A few figures are, however, necessary. Mr. Watson, the able Secretary of the Miners' Permanent Benefit Fund, has given me the following statistics as to the proportionate frequency of eye accidents among miners to other accidents. In all these accidents the miners have been rendered unfitted from continuing their work, at least temporarily. The figures are for 15 years, arranged in periods of 5 years. It is curious to see the proportion of 5 per cent. coming out practically the same for each period. The number of non-fatal accidents dealt with is 48,262.

Period	Number of Accidents	Number to Eye	Percentage
1884 to 1888	16,870	857	5·08
1889 to 1893	12,768	670	5·24
1894 to 1898	18,624	979	5·25
Total ...	48,262	2,506	5·19

The average yearly membership for each period was: 1884 to 1888, 22,410; 1889 to 1893, 17,876; and 1894 to 1898, 23,005.

Mr. Bridgeman, the secretary to the Equalised Druids Society, a benefit society embracing all classes of workmen, has given me other statistics. This society gives to

its members who are permanently incapaci-
ated from following their employment a
grant of £100. The number of cases of
all accidents in which this grant has been
made during the last five years is fifty-
seven, and out of that number the recipient
has obtained it seven times owing to eye
accidents. My own infirmary figures (which
Dr. Barker, my house-surgeon, has kindly
worked out for me) also testify to the large
number of eye accidents annually occurring
in the district of which I am more particu-
larly speaking. Of the last 2,554 patients
who have passed through my wards at the
Sheffield Royal Infirmary 2,038 were men
and 516 women. Of the 2,038 men 622
were admitted for accident, or 30·52 per cent.
This percentage has kept fairly uniform, but
at periods like the present of great trade
activity the ratio of accidents to other cases
admitted has gone up. Of 516 women only
36 were admitted for accidents, or 6·9 per

Iron and Steel Works.

It is necessary to exclude entirely from my remarks all reference to accidents caused in everyday civil life, much as could be said about their prevention. My observations have reference only to accidents met with in connection with labour and, as already stated, they have a more particular bearing upon accidents occasioned by such trades as are carried on in Sheffield and the surrounding districts. Iron and steel works are no longer confined to one district, but are located in many different parts, and there are many other industries to which the general principles here set forth will equally apply.

The important part occupation bears to the number of eye accidents is well illustrated by my infirmary statistics, which I have already quoted. The men not only exceeded the women very largely in actual numbers, but still more so by percentage, this latter being six times as great as for the women.

Fig. 1.—Grinders: Edge tool grinding.

GRINDERS.

In many trades associated with iron and steel in all its varieties, small foreign bodies are very prone to become lodged in the workmen's corneæ. I take as an example the grinders. In the course of the day a grinder may get several foreign bodies fixed in his cornea, or days may elapse without such a mishap. " Mote " is the name popularly given by the workmen to these particles in their eyes. If the cornea of a grinder be carefully examined with a magnifying glass it will not infrequently be found to be studded over with minute nebulæ ; though, therefore, the damage done by each "mote " may often not be serious, yet the frequent repetition, by dulling the cornea, will, in many cases, diminish the acuteness of vision. These particles may either be small fragments of stone, or, much more frequently, small portions of steel, or emery, which latter is largely used as a wheel for glazing cutlery and for other purposes.

A few observations about the grinders'
occupation will be of interest. I take works
where, at the time of my visit, besides outworkers, 700 men were engaged in making
pocket-knives and razors, &c. Of the two
varieties of grinding, it was at once evident
that the dry grinders were more exposed
to injury from foreign bodies than the wet
grinders. A grinder sits across his bench,
which he calls his "horse," and presses the
knife or razor blade on the stone. The wet
prevents the particles flying about a good
deal, but still a man's face becomes, as he
works, bespattered ; nevertheless, a wet
grinder would say that, compared with the
dry grinder, he seldom gets "motes" in his
eyes. In dry grinding the sparks fly freely,
and it is evident that particles, very minute,
of steel or stone are being projected about,
and it is the merest chance whether they hit
the man's eye or face or scatter about the
room. The fans, which it is well known
have for many years been required in the

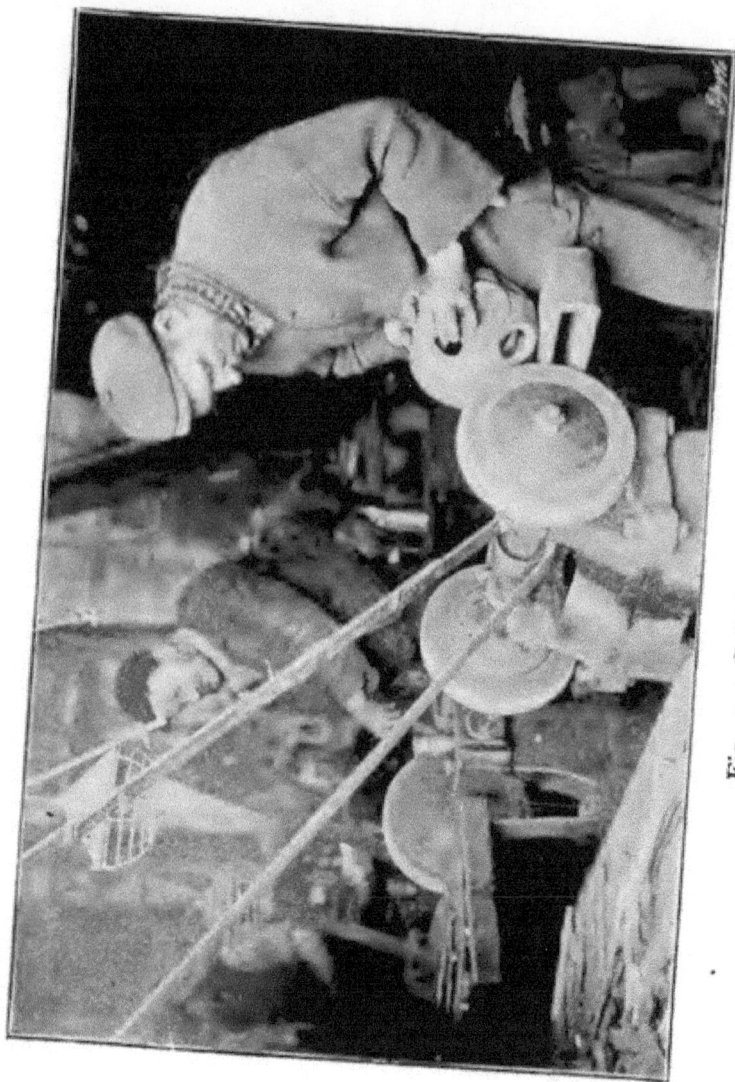

Fig. 2.—Grinding : Emery wheel.

grinding trade in consequence of its dele-
terious effects upon the health of the operatives, must be regarded as in some measure
a protective means. It is interesting to
observe the remarkable way a fan draws into
it the sparks and particles flying from the
wheel. There can be no question that the
grinder derives considerable immunity from
these "motes" by the employment of protective glasses. Several grinders whose ocular
condition necessitated the wearing of spectacles have admitted the protection they
afforded. If further testimony were needed
it can be found in the condition of the
glasses after use for some time by a grinder.
I have here a concave glass[1] which a grinder
used at his work. It will be noticed that
its surfaces are studded all over by small
dots ; both surfaces in this instance, because
the side of the glass had been changed ; so

---

[1] The other lens in the spectales had not long been
renewed, and was therefore less studded with the
dots referred to.

that when one surface had become studded the position of the glass in the frame had been reversed, and that which was next the eye had been placed externally, and had also in turn become studded with marks. This may be regarded as what occurs to the cornea in consequence of the repeated lodgment of particles.[1]

The nearer a man is to his work the greater is the chance of his being hit by one or more of these small particles. This suggests the proper correction of errors of refraction by the appropriate spectacles.

---

[1] After a time, in grinding blades a stone gets blunted, and its pores become filled with portions of hard steel. A process is then adopted for " cleaning " or " sharpening " the stone, and it is more than usually dangerous, because it liberates the embedded particles, as well as makes fresh ones. A bar of soft iron is held against the stone, in the opposite way to which the wheel is running. This is called " straddling." A great deal depends on the fineness of a stone, and accidents are much more common with new rough ones, as particles are much more likely to fly from them. The mode of preparing a

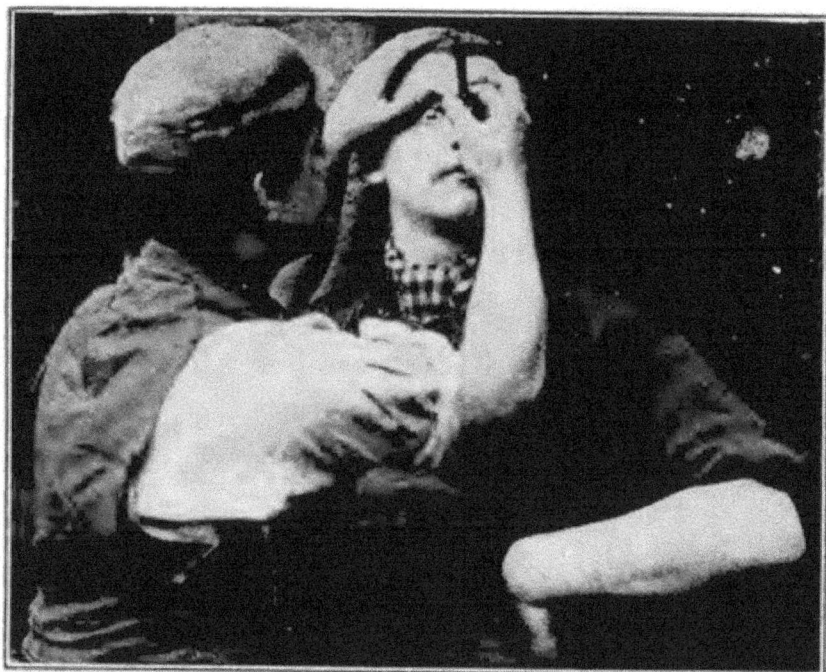

Fig. 3.—A workman removing a foreign body (mote) from a comrade's eye.

It must be admitted that in the great majority of instances the damage occasioned to the grinder or other operatives in which similar mishaps occur is not attended with serious results. To many, however, the immediate injury is serious either directly or by the ulceration that ensues. There is another way also in which injury results. A man once said to me, pointing to his damaged finger, "This would not have happened if something had not got into my eye, because I could not see my finger on the circular saw." He was an ivory cutter.

---

new stone is a dangerous one; it is called "racing." The stone is put on the band and it is turned the wrong way, and then a man holds a piece of steel against it to smooth its surface. The dust is tremendous, and splinters of stone fly in all directions. The men cover their mouths to keep the dust out, but they do not generally use any protective over their eyes. It is during this process that stones are apt to crack and fly, and that accidents occur entailing serious results to life and limb. An emery wheel is used for glazing, and here again much depends on whether the wheel is new and rough or not.

"The Mote Remover."

In the various trades in which iron and steel are used the operatives are liable, though to a less degree than the grinder, to get these "motes" into their eyes. Many workmen are skilful in removing "motes" from their comrades' eyes. In all the large works there are men who have a reputation in this way. It is, besides, not an infrequent sight, even in the streets, to see a man with his head pressed against a wall, and a fellow workman endeavouring to remove a foreign body from his eye. The number of foreign bodies some of these men remove in the course of a day is very large. One man, a time-keeper at works where 1,000 men, besides outworkers, were employed, told me he had for fifteen years at least been recog-nised as a skilful remover of "motes." Sometimes he had extracted a score or more a day; sometimes the number was much less, but he had not for many years passed a day without at least one case. He was not the only man with a reputation at these works, for grinders, and others, also removed motes. He used a lancet with the end blunted. It was quite clean, and he kept it so by either putting it on a strop or a wet stone. For the same purpose other men would use a lancet, pocket-knife blade, or even a pin, &c. The knife would very possibly be the same used at the man's lunch, and the pin not infrequently is put into the mouth to wet it before being used, and is very likely kept stuck in the man's waistcoat, where it comes in contact with the dirt collected in the clothing.

INFECTION OF CORNEAL TUMOURS.

Without doubt, in many instances, these motes are skilfully removed; in others there is a good deal of bungling. The instruments generally used are unsuitable. Not infrequently cases come under observation in which sloughing corneal ulcers have resulted

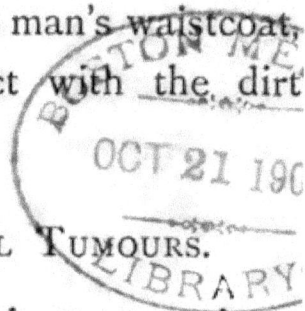

from the efforts made to remove a "mote."
It seemed to me not unlikely that a septic
condition was set up in consequence of the
uncleanly instruments which were so often
employed.   Dr. Shennan, of Edinburgh,
kindly undertook a bacteriological examina-
tion of some of these instruments for me.
I collected 22 tools used by different men;
17 of these were got together for me by the
skilful time-keeper I have already mentioned;
the other 5 were obtained at a large engine
works.   The photograph shows these instruments except two pins which were acciden-
tally omitted.   Some of the tools were fairly
clean, but others were in a dirty state; two
or three were magnetic.   Dr. Shennan examined the majority of these tools.   Taking
all in all, he found nothing pathogenic
excepting the staphylococcus pyogenes albus,
whose virulence is comparatively slight.
The other organisms found were chiefly
sarcinæ of the more common varieties, and
bacilli of the "subtilis" (hay bacillus) group.

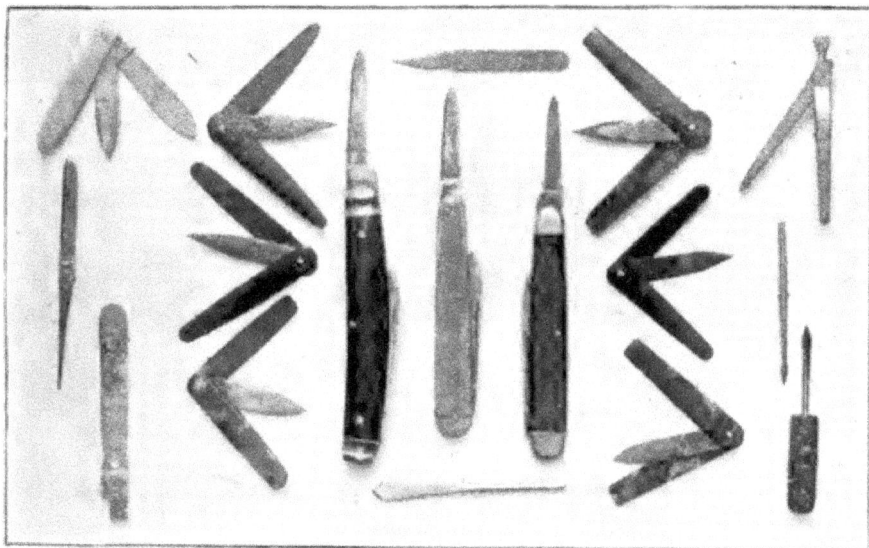

Fig. 4. —A series of instruments used by men for removing foreign
bodies from the cornea.

He failed to obtain cultures from one only of the instruments, and the greatest number of separate colonies were obtained from a common white pin. The presence of organisms had apparently no special relation to the cleanness or rustiness of the tools. They were present on the brightest steel, whereas the instrument from which he failed to obtain cultures was a lancet, the blade of which was slightly but decidedly rusty. Still, as he remarks, one cannot lay much stress on such an occurrence, as it might be accidental.

Of course, there are many sources besides these tools by which a corneal wound may become septic. But good should result if something cleanly and more suitable could be made accessible to these removers of "motes." A member of a large firm has promised to provide the men at his works enjoying reputations for removing motes among his employees with such instruments as, in my opinion, would be suitable. After considering different means which could be adopted for sterilising spuds, the simplest device seemed to me to be to provide a case containing a couple of iridium platinum blunt pointed spuds, together with a small spirit lamp, with directions printed on the inside of the case saying that before use the extremity of the spud should be heated in the flame of the spirit lamp, or, if this be not accessible, in a gas or other flame which may be at hand. Messrs. Down Brothers have made such cases on my suggestion.

IRON AND STEEL WORKERS.

By far the most serious eye accidents happen to men engaged in working iron or steel. The following figures exhibit this in a very lurid light.

STEEL AND IRON.

Steel and iron splinters, rivet, chips, pieces of drill, file, wire, &c. 173
Nail 5

BURNS.

Metal sparks, flashes, &c. 43

Lime 8
Gas explosion ... 1
Ammonia 1
Gunpowder ... 4
Cinder 1
Poker 1

MISCELLANEOUS.
Dynamite, and dynamite explosion, and cartridge 6

Wood, sticks, and peggy	...	...	13	
Hook	...	...	1	
Knife	...	...	8	
Glass, soda water bottles, &c.	...	15	
Pick	...	...	5	
Stone	...	...	24	
Fork	...	...	7	
Pin	...	...	1	
Fist	...	...	3	
Branch of tree	...	1	
Crane handle	...	1	
Cork	...	...	2	
Cinder	...	...	4	
Coal	...	...	11	
Straw	...	...	1	
Cat's claw	...	...	1
Sand	...	...	1
Ball	...	...	1
Pen	...	...	2
Firework	...	...	1
Boiling oil	...	...	1
Tin	...	...	2
Band strap	...	2
Band buckle	...	2
Chain	...	...	1
Kick	...	...	1
Brick	...	...	1
Thorn	...	...	2
Elastic, piece of	...	1
Total	...	359

Out of this total of 359 eye accidents to males, taken from the records for this purpose consecutively, which were so serious as to require admission to my wards at the Sheffield Royal Infirmary, no fewer than 173 were caused by iron or steel, including splinters of iron or steel, pieces of rivet, of drill, wire, and many other means associated with the iron and steel trades. There were also 43 due to burns from molten metal, sparks, flashes, &c. I am not sure, also, whether to the former number should not be added 5 put down as caused by nails, as

most, if not all, of them would have occurred
to iron or steel workers.

"CHIPPERS" OR "FETTLERS."

The opportunity for the infliction of severe
injuries to iron and steel workers is multitu-
dinous. They occur in all branches of the
trade, in the lighter iron and steel industries
as well as in the heavy trades, where armour
plates are made and heavy castings of
scores of tons. A very large proportion of
the accidents are occasioned by what is
called "chipping" and "fettling." "Dress-
ing" is the name given in some parts to this
process. This work consists in chipping the
rough edges from steel and iron castings,
ingots, and all kinds of iron and steel work,
and, among other things, even the large
armour plates. I append a short list, which
a patient, himself badly injured when passing
men engaged in "chipping," has furnished
me with, of the articles in connection with
locomotive work, as an example, of the

Fig. 5.—Men engaged in chipping.

number of objects in which "chipping" is required.

Rolled Steel.—The frame plates for locomotives require chipping for the hornblocks or axlebox slides.

Cast Iron.—The saddle casting which is placed under the front end of the boiler and between the frames to which it is bolted. This casting wants chipping to the radius of the boiler.

Cast Iron.—The dragbox, which is placed between the frames at the back of the engine.

Cast Steel.—The hornblocks or axlebox slides sometimes require chipping and facing. The platform plates (of mild steel), which are the plates round the engine, require chipping on the edges.

Steel.—The boiler has to be chipped and faced in various places for the mountings.

Cast Iron.—The chimney, if a casting, requires chipping and bedding to the boiler.

Steel.—The clothing plates for the boiler.

These plates are rolled to the radius of the boiler, and then laid over a block, and pieces from 1 to 4 or 5 inches in diameter are chipped out to clear different mountings, &c., on the boiler.

Castings of either iron, steel, or brass are the most dangerous to work upon, because the chippings fly about on account of the metal being brittle. It is very dangerous chipping castings in the corners or where the chipping strikes the metal and rebounds. Chippings from the castings are about $\frac{1}{4}$ inch to $\frac{3}{4}$ inch long and very sharp. When chipping thin plates on the edges, the chippings are sometimes 1, 2, or 3 inches long before they break off. All castings are "fettled" at the foundry, that is, the runners are cut off, and the places where the metal has run at the joint of the moulding boxes are trimmed off.

Whatever be the special kind of metal or steel to be fettled, the manner in which it is done is practically the same. A hammer

Fig. 6.—Chipping: Man wearing protectors.

and chisel or sate are used, and with these the roughnesses are removed. Frequently, also, whilst one man places the chisel, another, or even two others, will use a hammer and are called "strikers." I understand that at works where, say, 1,000 men are employed, 200 or more will be occupied more or less in "chipping." Many men are frequently working close to each other, so that the danger is not only to the worker himself but to those around. Passers-by are by no means infrequently the victims, and many blinded in this way have come under my notice. The chipper himself is often hit by the rebound of the splinter after it has struck perhaps the narrow angle of steel or iron upon which he may have been working, or some other object. It must be recollected, also, that in the process spoken of the danger is not merely from the iron or steel which is being operated upon, but there are three other places from which splinters may be, and actually are, given off and cause injury—

namely, the hammer head, the chisel head, and the chisel point.

It is obvious that men engaged in work which causes the splinters to fly about so freely should be so arranged as not to be chipping against their fellow workmen, or in a direction from which passers-by may approach. This is managed in some works by getting the men to chip against a wall, though not too close to it, or again by interposing a canvas screen between sets of workmen.

The sizes of the splinters spoken of vary from the most minute to others measuring some inches in length, and they may be thick or thin. The injury inflicted differs, of course, in accordance with the size of the missile and the force with which it is projected. The small fragments may be thrown off with such velocity that they penetrate the eyeball and become embedded in its interior, in some instances passing through the eyelid before reaching the globe. The destruction to sight in this way is very large. I have myself

Fig. 7.—Chipping against a screen. Men wearing protectors.

removed from the globe, with the electro-
magnet, I believe more than 200 fragments
of steel and iron, mostly projected into the
eye in the manner I have just mentioned.
I have arranged in a case for inspection 117
of these splinters. Though all are comparatively small, there is a great variety in size.
One is no heavier than 0.0015 gr. ; there are
several as light as 0·0030 gr. and 0·0046 gr.
The largest weighs 36 grs., and there are
two others 12 and 9 grs respectively. This
is not the place to refer to the results of the
extraction of this number of foreign bodies
with the electro-magnet, and I content myself
with saying that many eyes have been saved
by its employment which otherwise would
have been hopelessly lost. The large chippings may occasion such extensive wounds
of the eyeball that the eye is at once irre-
parably damaged, or is so injured that re-
moval of the globe will subsequently be
necessitated.

It has been my sad fortune on more than

one occasion to have a workman under treat-
ment in consequence of a severe injury caused
by a large or small splinter of steel or iron,
and in which loss of sight has resulted, who
has afterwards returned to his work and has
again come under my care with a similarly
disastrous accident to the remaining eye.
In one instance, which I well remember, a
very few weeks only had elapsed between
the two accidents.

THE PNEUMATIC CHIPPER.

Besides the manner of arranging the men
at their work to which reference has already
been made, and the means to be recom-
mended later for adoption by the workmen,
it is fortunate that the dangers of chipping
may be avoided by adopting a pneumatic
chipper. My friend, Mr. Bernard A. Firth,
of the well-known firm of Messrs. Thomas
Firth & Sons, has kindly given me the
following note on this method. He says :—

" For chipping ingots of steel or anything

Fig. 8.— Pneumatic Chipper.

that is of a comparatively soft nature—that is, not very brittle—we have almost entirely given up the use of hand chipping by sate and hammer, and adopted a pneumatic chipper. The advantage of this is that the pieces do not fly from a chisel, but merely curl up and fall over. In cases where this cannot be used the blocks to be chipped should be ranged near the walls inside the building, with the men facing the wall, so that they do not chip towards each other. A certain number of accidents occur among men known as fettlers, who are employed in chipping the rough edges from steel castings. Up to the present these pneumatic tools have not been a success for this work, but possibly, in due time, some improvement will be made which will enable them to be used."

I visited these works, and there saw the pneumatic chipper at work on a large casting. It has the advantage of accomplishing in one hour what, by hand, would take six or seven hours. It certainly prevents the flying about of splinters in a remarkable manner. It is more like using a cheese-scoop in a fairly soft cheese than running any tool over hard steel.

Injuries from Molten Metal.

Another class of severe injuries which are of common occurrence are burns from molten metal. Sparks and flashes fly about freely in almost every instance that molten metal is run into the moulds; but in some, of course, more so than in others, and the portions given off vary much in size. Injuries caused in this manner were no fewer than 43 out of the 359 consecutive accidents in males already mentioned as admitted into the Royal Infirmary under my care. In the forgings, also, great or small, when the iron or steel is being hammered either by hand, or, in the case of larger castings, by a steam or hydraulic hammer, portions are given off from the glowing metal, and those working, and the bystanders, are exposed to danger of

Fig. 9.—Crucible steel melting.

burns. Injuries inflicted by molten metal are very serious, owing to the immediate destruction of tissues or from the later results.

Preventive Measures.

The knowledge which I have acquired from contact with working men who have been injured, and visits from time to time to many of the principal works, has long since satisfied me that much of the destructive injury to sight, which is constantly coming under one's observation is preventable, and that means can and should be adopted to lessen the risks to sight which are at present associated with such important industries. My experience has shown me that there is less difficulty in enlisting the support of the employers than the assent of the men to adopt precautionary measures. With the Compensation Act in force I have no doubt the assistance of the masters will even be stimulated. It only recently came to my knowledge that one firm had made it compulsory on men engaged in "chipping," "fettling," "turning," and other work in which iron and steel splinters were liable to fly off and endanger sight, to wear protectors, which were provided at the expense of the firm. This decision was taken in consequence of a workman being blinded by a "chipping," and it is easy to understand, while creditable to their humanity, how such a course, if it tends to preventing loss of sight, is likely to be a considerable pecuniary saving.

Eye Protectors.

In considering what kind of protectors men should use, certain points must be borne in mind. The cost must be very moderate, and any covering over the eyes should interfere with the sight as little as possible, if at all. Among iron-workers, glass is practically out of the question. Even thick rock crystal, which has been suggested for some kinds of work in consequence of its thickness and

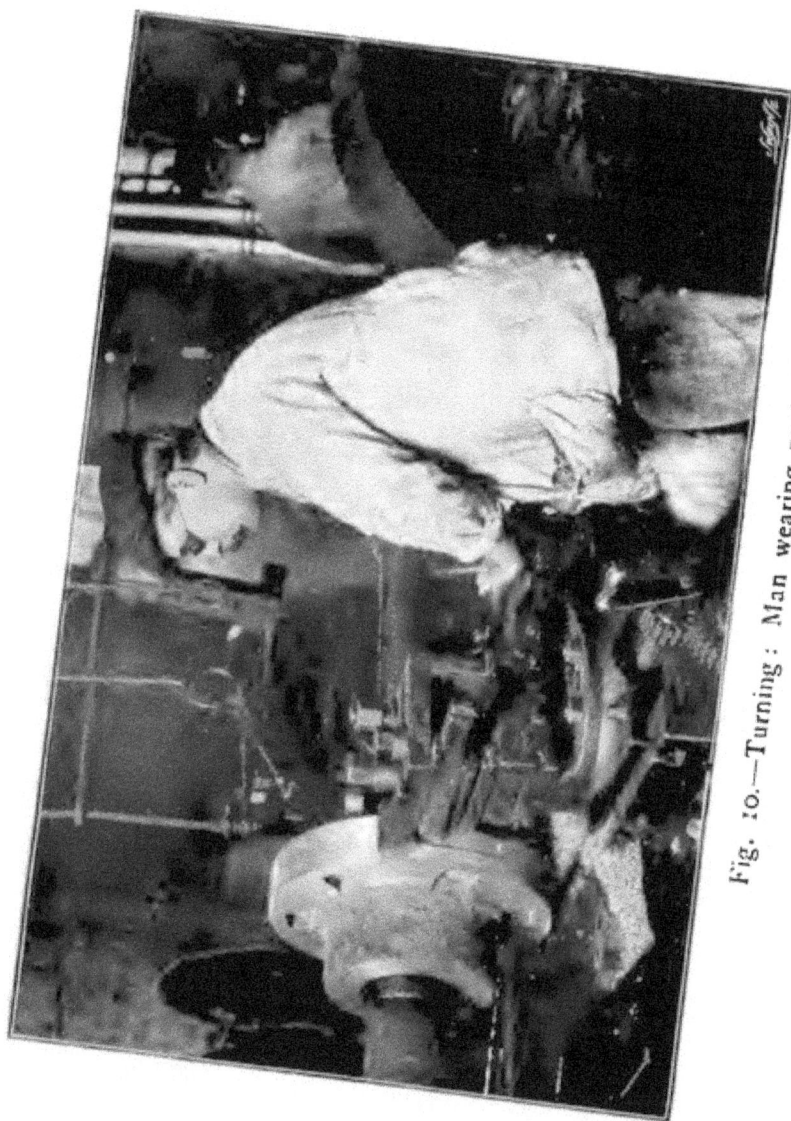

Fig. 10.—Turning : Man wearing protectors.

peculiar manner of fracture, would hardly do. There are a variety of protectors in the market, but they have been little used.

Gauze wire, fitting close to the eye like a cup and attached to the head by a string, is employed by stonebreakers and in some ironworks. I have photographs of men working with these protectors over their eyes. They are taken from men actually engaged at their daily occupation at the works (*vide* Figs. 6, 7, and 10) to which reference has already been made, where the men are practically compelled to adopt protectors. They complain of them as being hot and interfering with sight, but there is no question that such protectors do afford considerable immunity from accident. Another practical point about protectors is that they should not be liable to rust. For this reason galvanised iron wire, or, better, aluminium wire, is of service. The mesh should be sufficiently strong and fine, and sufficiently close to prevent, as far as possible, even small chippings passing through it, and yet to interfere with sight as little as need be. I have had experiments made by allowing men engaged in fettling to "chip" against wire gauze which has been suspended for the purpose, to ascertain how far a mesh answered before deciding to adopt a given size. Then, acting on a suggestion derived from Bronner's shield for operation cases, I have had this netting made[1] into protectors which cover the eyes and the adjacent parts. The portion over each eye is bulged forward so as to allow very free play to the eye underneath, and the convex surface will be a greater protection than one merely flat would be. I have supplied workmen with these protectors, who have used them for chipping, steel melting, and other dangerous iron and steel work. I learn that they are regarded as satisfactory, that they answer their purpose well as protectors,

---

[1] By Messrs. Priest & Ashmore, opticians, Sheffield.

Fig. 11.—Protectors (Author's pattern).

and that the interference with sight is very little.[1]

CONCLUSIONS.

I would sum up my suggestions as to the means for protection as follows :

(1) The grinder will find that large glasses made of plain glass, or, indeed, his own spectacles, should his refraction require their use, will afford great protection. Or he may use other protectors made with glass in front and gauze surrounding it.

(2) The use of protectors should be compulsory for those workers in iron or steel whose employment renders them liable to be injured by iron or steel splinters, or are exposed to danger from molten metal.

The gauze eye shield I have described will, I believe, answer the purpose well. The cost is low, and it is worth the employers' while to supply their men with them.

---

[1] The interference to sight is not greater than that occasioned by many ladies' veils.

Other means to be adopted are :

(*a*) The use of a pneumatic chipper whenever practicable ;

(*b*) The proper arranging of the men at their work ; and

(*c*) The use of screens so as to avoid injury to their fellow-workmen and to passers-by.

It is my belief that a consideration of the facts I have advanced will lead to the conviction I have myself long held, that very many eye accidents associated with trades are preventable, and to the view that, such being the case, preventive means should be adopted.